WORLD OF WONDER

Published by Creative Education
123 South Broad Street
Mankato, Minnesota 56001

Creative Education is an imprint of
The Creative Company.

Art direction by Rita Marshall
Design by The Design Lab
Photographs by Affordable Photo Stock (Francis & Donna
Caldwell), Peter Arnold (K. Schafer, Alain Torterotot), Corbis
(Gary W. Carter, Macduff Everton, Darrell Gulin, Steve
Kaufman), The Image Finders (Bill Leaman), JLM Visuals
(Richard P. Jacobs), Robert McCaw, BillMARCHEL.com, Tom
Myers, Seapics.com (Saul Gonor, Amos Nachoum, Robert L.
Pitman), Tom Stack & Associates (Jeff Foott, Thomas Kitchin,
Joe McDonald, Charles Palek, Milton Rand, Mark Allen
Stack, Spencer Swanger)

Copyright © 2003 Creative Education.
International copyrights reserved in all countries.
No part of this book may be reproduced in any form without
written permission from the publisher.

Library of Congress Cataloging-in-Publication Data

Hoff, Mary King.
Coping with cold / by Mary Hoff.
p. cm. – (World of wonder)
Summary: Describes the variety of ways in which creatures
cope with a cold environment, from antifreeze created by
trees to the hibernation of squirrels.
ISBN 1-58341-239-5
1. Cold adaptation–Juvenile literature. [1. Cold adaptation.]
I. Title.

QH543.2 H64 2002
591.4'2–dc21   2001047886

First Edition

9 8 7 6 5 4 3 2 1

cover & page 1: frost on a dragonfly
page 2: a bird in winter
page 3: a sleeping polar bear

Creative Education presents

WORLD OF WONDER

COPING WITH COLD

BY MARY HOFF

Trees with antifreeze 🌿 Mammals with extra-furry coats 🐿 Frogs that bury themselves in mud ❄ The world is full of living things that have special structures or behaviors that help them survive in cold times and places.

🌎

SUCH STRUCTURES AND BEHAVIORS are called **adaptations**. Some help creatures make or absorb heat, or preserve the warmth they already have. Others enable creatures to get by without much heat. Either way, they allow plants and animals to live in frigid conditions that might otherwise kill them.

Many Canadian geese migrate before winter

NATURE NOTE: *Frogs can raise their body temperature more than 18 °F (10 °C) above the air temperature by basking.*

SUN AND SHIVERS

In early spring, turtles clamber onto a log. They lie there sleepily as the sun beats down. They are basking—warming themselves with solar energy.

☀ Sunlight is a source of warmth for many animals. Frogs, toads, snakes, and other **cold-blooded** animals need it to keep going in cool weather. But **warm-blooded** animals such as mammals and birds heat their bodies with sunlight, too. A polar bear's fur has see-through tubes that conduct sunlight

toward its black skin. Because dark colors absorb light well, the skin warms quickly.

 The heat that warm-blooded animals make by **metabolizing** helps keep them warm. But when it's really cold, they may need to increase the activity of their muscles. Sometimes muscles start contracting and relaxing rapidly just to make heat. This is called shivering. Both mammals and birds shiver.

FAT, FEATHERS, AND FUR

As winter approaches, squirrels eat more. This extra food adds a layer of fat beneath their skin. Fat is a kind of **insulation**. It acts like a blanket, helping to trap body heat.

Fur and feathers also slow the escape of body heat. They do so by trapping air, which is a good insulator, near the body. Chickadees and other birds that live in

NATURE NOTE: *The coldest known air temperature on Earth, recorded at an Antarctic research station on July 21, 1983, is -128.6 °F (-89.2 °C).*

Walruses have a thick layer of fat for insulation

cold places grow more feathers just before winter. Some birds, including ducks and other water-fowl, have a layer of tiny, fluffy feathers called down that adds extra insulation. Squirrels, foxes, and other mammals grow more fur as winter approaches.

✷ When it's very cold outside, an animal's body may activate muscles that make its fur or feathers stand up, increasing the thickness of the insulating air. In humans this is called "goose bumps" or "goose flesh." Since we don't have fur, goose

NATURE NOTE: *Chickadees sometimes lower their body temperature a bit so they don't need to use as much energy to stay warm.*

A chickadee grows extra feathers for winter

bumps don't do much to insulate us, but they do remind us to put on a jacket or do something else to get warm.

❄ Birds such as grouse and ptarmigans use snow as insulation. They dive or burrow beneath the surface to stay warm. If they go deep enough, they might run into signs of other animals. That's because snow also shelters mice, shrews, lemmings, and weasels. These creatures stay warm by living in tunnels in the fluffy, insulated **subnivean** world.

NATURE NOTE: *Furry insulation makes it possible for arctic foxes to sleep comfortably when the air temperature is -40 °F (-40 °C).*

A fox's thick fur protects it from the cold

BEING BIG

Some penguins live close to the South Pole. Others live near the equator. The polar penguins are bigger than the equatorial ones. Scientists think they survive better than small ones would because being big helps in staying warm.

Why does being bigger keep an animal warmer? Larger objects have relatively more insides and less surface area than smaller objects. The larger an animal is, the more heat-producing body it has relative to the amount of heat-losing skin.

A variation on being big is to crowd together—to be big as a group. White-footed mice, tundra voles, and other small rodents gather with others of their species so the heat from one can warm the others. Quail, bees, penguins, and muskoxen huddle together for warmth. Beavers share body heat by living together in an insulated structure made of mud, branches, logs, and stones.

NATURE NOTE: *The blubber (fat) in a bowhead whale can be up to 12 inches (30 cm) thick. This helps protect it from the ocean's cold water.*

Emperor penguins huddle for warmth

TAKING CARE OF TOES

For animals that stay active in cold weather, legs and flippers pose a special problem. They have a lot of surface area, so they have the potential to lose a lot of heat.

 Many animals that live in cold conditions have adaptations that reduce the loss of heat from these body parts. The rock ptarmigan, which lives in Greenland, has feathers on its feet that pro-

NATURE NOTE: *Countercurrent heat exchange cools the blood flowing into the legs of the arctic tern so much that its feet stay close to 32 °F (0 °C)—the freezing point of water!*

Feathers keep rock ptarmigans' feet warm

14

tect its toes from cold. Other animals have poor circulation in their legs. Because blood carries body heat, the less blood in these body parts, the less heat they lose. ℘ Blood vessels in the legs of penguins and some wading birds are arranged in a way that minimizes heat loss. **Veins** and **arteries** run close to each other inside the legs. As arteries send warm blood out to the toes, the cold blood in the veins picks up some of the heat they carry and transports it back to the body. This reduces the amount of heat lost. This process is called countercurrent heat exchange.

Penguins' squat shape reduces surface heat loss

A duck's feet can tolerate very cold water

THE BIG SLEEP

Some animals cope with cold by migrating to a warmer climate. Others go to sleep. As winter approaches in the northern tundra, the ground squirrel's body begins to shut down. Its heartbeat drops from 200 beats to 10 beats or less per minute. Its breathing slows, and its body temperature drops.

The ground squirrel is **hibernating**. The changes in its body allow it to survive cold weather without having to find lots of food. Other warm-blooded animals that hibernate include bats, marmots,

NATURE NOTE: *Black bears sleep a lot in winter, but most scientists don't consider them true hibernators. Their body temperature doesn't drop much, and they wake easily.*

chipmunks, and hedgehogs.

Cold-blooded animals hibernate, too. Snakes seek shelter in hiding places in the soil or in openings in rocks or stumps. Toads and frogs on land dig a hole, bury themselves in leaves, or hide in a crack in a log or rock. Frogs that live much of their lives in water spend winter at the bottom of a pond or creek. Turtles also burrow into the mud. Hibernation slows their bodies down so much that they need little oxygen. They can get the oxygen they need from the mud around them.

NATURE NOTE: *A bat may lose one-third of its body weight while hibernating.*

Like many mammals, bats sleep through winter

PLANT SOLUTIONS

A big challenge for plants in cold weather is a lack of water, since cold air doesn't hold much moisture, and liquid water turns to ice. To minimize water loss, needleleaf trees have small, waxy leaves. Broadleaf trees such as maple and oak drop their leaves in autumn. If these trees kept their big, flat leaves in winter, when roots have a hard time drawing water from the cold earth, the leaves would lose so much water that the trees could dry out.

The waxy needles of a pine tree conserve water

    Cold also poses a problem for plants' cells—the tiny, liquid-filled structures that make up living things and keep them going. If the liquid freezes inside cells, it could cause them to break open, much as a sealed beverage container breaks open when it freezes. Some plant species avoid this problem by leaving hard, dry seeds that survive the winter. **Perennials** such as grasses concentrate their growing power in the part of the plant that is protected beneath the ground.

Trees and other plants that keep living parts above ground in very cold weather go through a process in the fall called hardening. They begin to dry out. As the water inside their cells decreases, the temperature at which the cells freeze decreases, too.

Oaks, elms, and other cold-tolerant plants also take advantage of a tactic known as super-

NATURE NOTE: *Hardening explains why plants may die from a late or early frost, even though they easily tolerate cold in winter.*

Maple trees dry out to survive the winter

cooling. Although the freezing temperature of water is 32 °F (0 °C), ice won't form at that temperature unless there is a bit of dust or other material in the water to start the process. By keeping the insides of their cells free of such bits, plants can keep water liquid inside of them when it's as cold as -40 °F (-40 °C)!

NATURE NOTE: *Aspen trees have chlorophyll (the pigment in leaves that helps plants photosynthesize, or make food using the sun's energy) in their bark. This allows them to make food even in winter.*

Aspens drop their leaves to conserve water

Wood frogs can survive extreme temperatures

ANIMAL ANTIFREEZE

Cold-blooded animals use some amazing strategies to prevent problems inside their bodies when temperatures plunge below the freezing point of water. The wood frog moves water from inside its cells into the spaces between them when it gets cold. It also produces a kind of antifreeze. When ice does form inside the frog's body, it forms outside of the animal's vital organs. A wood frog can survive even after most of its body freezes solid.

Some insects make a kind of antifreeze, too. They also move water from inside to outside their cells and rely on supercooling to prevent damage to their cells caused by ice crystals. The process of supercooling allows insects in the far North to keep their cells free of ice crystals at temperatures as low as -22 °F (-30 °C).

NATURE NOTE: *Some insects go through a period in winter during which they don't move or grow. Scientists call this state diapause.*

Ladybugs become inactive during cold periods

28

A PERFECT FIT

From blubbery whales to frogs with antifreeze, the world is full of creatures with adaptations to tolerate cold. These adaptations make it possible for life to exist in places that would otherwise be empty. They add threads of richness and variety to the multicolored fabric of life on Earth.

❄ Adaptations to cold are just some examples of the many ways in which living things fit amazingly well into the world around them. When humans make changes to the natural world, we also risk destroying that "perfect fit." By considering the impact our actions have on the environment and its wild creatures, we can help ensure the future health and beauty of this amazing world, this world of wonder.

NATURE NOTE: *In a beaver's tail, arteries and veins branch out to make a web known as a rete mirable. This improves counter-current heat exchange.*

NATURE NOTE: *Some birds hide their bills in their feathers to protect them from the cold.*

A white-tailed jackrabbit burrows in the snow

WORDS TO KNOW

Traits that help a living thing survive or reproduce under the particular conditions in which it lives are called **adaptations**.

Arteries *are blood vessels that carry blood from the heart.*

In **cold-blooded** *animals, body temperature changes as the temperature of the surroundings changes.*

Some animals endure cold weather by **hibernating**—*slowing down their body processes and going into a state that seems like sleep.*

Insulation *is a layer, such as feathers, fat, or fur, that slows or limits the loss of body heat.*

When an organism is **metabolizing**, *it is carrying out processes within its body that provide the heat and energy it needs to live.*

Perennials *are plants that live more than two years.*

Subnivean *animals live beneath the snow.*

Veins *are blood vessels that carry blood from the body to the heart.*

In **warm-blooded** *animals, body temperature is determined internally rather than by the temperature of the surroundings.*

INDEX